Marcelo Henrique Santos
Bruno Santos
Jose Ricardo Marques

Interactive Digital Television in Brazil

Marcelo Henrique Santos
Bruno Santos
Jose Ricardo Marques

Interactive Digital Television in Brazil

Brazil's Interactive Digital TV technology - SBTVD - Brazilian Digital TV System

ScienciaScripts

Imprint

Cover image: www.ingimage.com

This book is a translation from the original published under ISBN 978-613-9-61677-0.

Publisher:
Sciencia Scripts
is a trademark of
Dodo Books Indian Ocean Ltd. and OmniScriptum S.R.L publishing group

120 High Road, East Finchley, London, N2 9ED, United Kingdom
Str. Armeneasca 28/1, office 1, Chisinau MD-2012, Republic of Moldova, Europe
Printed at: see last page
ISBN: 978-620-7-88721-7

Table of contents:

Chapter 1

1. INTRODUCTION

On May 11, 1928, in New York, WGY - *Wireless General Electric in Schenectady made* the world's first television signal transmission. At the time, the devices were very simple and were often radios adapted with screens. A few years later, with the technological advances achieved mainly during the Second World War, the first analog high-definition signal was transmitted.

As early as the 1950s, the United States saw the emergence of color signal transmission, along with the arrival of the first color televisions. There was also an adaptation of non-color equipment during this period.

Our country saw its first TV broadcast in 1948, which aired a soccer match in Minas Gerais. Two years later, in Sao Paulo, the first commercial signal was broadcast in Brazil by the now defunct TV Tupi, making it the 4th country to have this technology, in addition to the United States, France and England.

Over the years, and with the creation of other television stations in the country, the signal spread to other regions. In the 1960s, the space race helped send the first television broadcasting satellites into space. The result was satellite broadcasting, making it easier to access the signal, regardless of its location, and improving quality.

Other elements began to emerge, such as cable operators offering their consumers access to broadcasters from other countries. Until on December 2, 2007, in Sao Paulo, the first digital television signal was broadcast in the country.

Digital Television in Brazil is still being implemented, so it is necessary to know about this technology in order to understand the changes that will take place in the country's main means of communication.

1.1. OBJECTIVE

The aim of this work is to present the Interactive Digital TV technology being implemented in Brazil, the SBTVD - Brazilian Digital TV System, to discuss the different technologies adopted by various countries and to develop a software application that promotes the interactivity proposed by this new technology.

1.2. METHODOLOGY

The development methodology for this work consists of the stages described below.

In stage 1, a bibliographic survey was carried out, using articles, books and publications, on SBTVD technology, showing the structure of this system, the reason for choosing it and the real objective of implementing this system in Brazil. The result was technical information and more in-depth concepts in the area of digital TV technology.

Stage 2 involved a series of interviews with professionals working in the field and/or studying Interactive Digital TV in order to draw a parallel between theory and market practice. The result was a better understanding of this technology in a practical way, verifying the changes that are being made to adapt to this new system.

Stage 3 saw the development of an application that simulates Interactive Digital TV. The result is a real example of the usability of this new technology.

1.3. WORK STRUCTURE

The rest of this paper is structured as follows:

- Chapter 2: **DIGITAL TV technology**, which defines the three main technologies currently in use: the American system, the European system and the Japanese system, and presents concepts on the subject;
- Chapter 3: **Brazilian Digital Television System,** sao the reasons for choosing this technology and why it is so important in the globalized world;
- Chapter 4: **Professionals' Reports**, presents a series of interviews with professionals working in the market;
- Chapter 5: **Example of an Interactive application**, presenting a system model that simulates Brazil's Interactive Digital TV technology;
- Chapter 6: **Conclusion**, where the conclusions of the work and proposals for its continuation are presented.

Chapter 2

2. DIGITAL TV TECHNOLOGY

Cruz (2008) states in his report that "[...] digitization can be considered the biggest technological change in the history of television. Even bigger than the transition from black and white to color."

Digital TV technology provides high-definition transmissions, the image line format in the Digital System is widescreen[1] . A widescreen image is one whose aspect ratio[2][3][3] is 16:9. In traditional televisions, the aspect ratio is 4:3.

While in the analog system the broadcaster can send only one program at a time, in Digital it is possible to send up to six programs simultaneously, allowing you to vary the programming or offer a richer experience, such as watching a soccer match from different cameras. This feature in sports coverage gives us the feeling that we are watching the event where it is happening. These features are exploited by Japanese public broadcaster NHK3 - Nippon Hoso Kyokai, in its soccer match broadcasts, as seen in Figure 1.

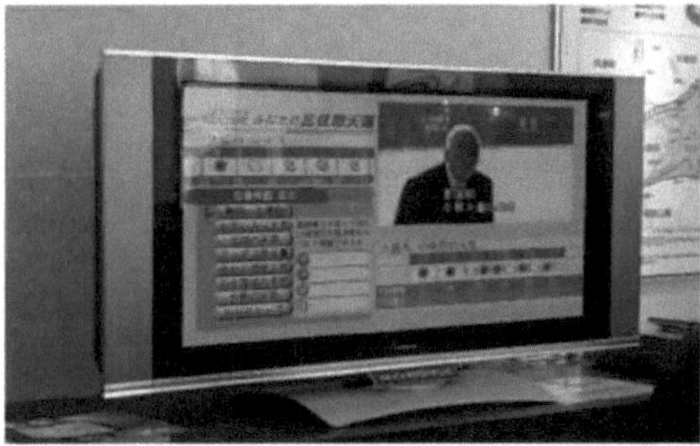

Figure 1 - Image of a digital TV broadcast by the Japanese network NHK

Interactivity allows users to take a variety of actions, such as voting in electronic polls, consulting the broadcasters' program guide, without using a telephone or computer. Despite these technological innovations, the most impressive are the audio and video transmissions that are being handled digitally, can be stored and can be accessed remotely and simultaneously by an unlimited number of people. This concept is already being used in the implementation of multimedia digital libraries, which are replacing conventional libraries by digitizing their collections[4] .

> Digital TV allows for high definition (an image six times better than the conventional one), multiprogramming (broadcasting several programs at the same time on a single channel), interactivity (Internet-like services on the device) and mobility (receiving free-to-air TV on cell phones and other mobile devices). (CRUZ, 2008)

With this technology, multimedia resources (audio and video) are immune to interference (noise during handling), because when transmitting digital information, small deformations or interference can be detected and corrected automatically by the receiver and converter, known in the market as a *Set-Top Box*[5] .

---

[1] The aspect ratio of a movie projection (and the format of the movie on your TV) is expressed as width divided by height.

[2] The aspect of the screen is determined by the ratio of the width of an image to its height.

[3] Japanese broadcaster's website - http://www.nhk.or.jp/

[4] The Multimedia Digital Library project provides the population with a large digitalized collection, democratizing access to the country's cultural assets via the Internet, which can be consulted at http://www.institutoembratel.org.br/.

[5] It's a term that describes a piece of equipment that connects to a television and an external signal source, and transforms this signal into content in a format that can be displayed on a screen.

2.1 COMPONENTS OF INTERACTIVE DIGITAL TV

According to DTV[6] (2008), digital TV technology "[...] transforms every tiny element of the scene and sound into a binary number made up only of zeros (0) and ones (1); it is the same technological language as computers".

In order for this interactive technology to work, a series of techniques have been developed over the years to reach its current stage. These techniques result in both the mode of transmission and the quality of the image.

Another important point is the issue of data compression, since many scenes can contain a lot of white space, causing signal loss or the appearance of pixels on the screen that interfere with the display of the image on high-definition television.

2.1.1 MEANS OF TRANSMISSION

According to DTV (2008), analog TV is a form of continuous image and sound. Today, you see images with blurred outlines, ghosts, noise, distortions in the color of people's skin, difficulty reading text and small numbers and, in addition, you hear unpleasant sound and often, even using stereo sound technology, the quality doesn't show results. pleasing to the consumer.

Digital signal transmission is called broadcasting[7]. There are currently four known means by which it is possible to realize and, consequently, use Digital TV resources: via Terrestrial, via Satellite, via Cable and via Internet.

Terrestrial transmission is carried out using radio frequency waves which require appropriate antennas and receivers for reception. This is probably the most eagerly awaited form of digital TV, according to Montez (2005), because the cost of implementation for viewers is low (compared to other means of transmission).

Satellite broadcasting[8] has been in use in Brazil since 1996 via pay-TV (companies such as *SKY, Century and DIRECTV*). This system allows people living in distant regions to receive the Digital signal and use all the proposed features. Technically, this system is called Digital C-band transmission.

Cable broadcasting uses conventional CATV - *Community Antenna Television* - cable networks to transmit digital signals, which reach subscribers' homes via pay-TV operators. Implemented in 2004 in large Brazilian cities such as Sao Paulo, Rio de Janeiro and Brasilia, this transmission resource is currently the most widespread in the world.

The last means of transmission mentioned in Montez's work (2005) is transmission via the Internet, also known as IPTV - *Internet Protocol Television*. This system uses high-speed connections (broadband) to transmit the digital signal. One of the advantages of this technology is the use of Internet resources (*web and e-mail*) alongside TV reception.

2.1.2 MULTIPLEXING AND MODULATION

According to Bernal (2005), multiplexing is the ability to transmit several signals, whether audio, video or data, sharing the same physical transport medium. This transmission can be carried out in two ways:

- FDM - *Frequency Division Multiplexing*: this technique allows a frequency channel to be divided into sub-channels using the same frequency band;
- TDM - *Time Division Multiplexing*: TDM allows the same channel to be divided into sub-channels and for them to work in different time slots.

Artuzi (2001) defines modulation and its importance as follows:

> A modulation process consists of modifying the format of electrical information with the aim of transmitting it with: the lowest possible power, the lowest possible distortion, ease of recovery of the original information, and the lowest possible cost.
>
> The process of amplitude modulation results in shifting the spectrum of the signal containing the information to a higher frequency in order to make it possible to transmit the resulting signal via electromagnetic waves, since higher frequencies allow efficient antennas to be built with reduced dimensions.

---

[6] Official SBTVD website, available at http://www.dtv.org.br.

[7] It's the sending of content (audio, video or data) from a service provider point to another point.

As for modulation, Montez (2005) mentions that the use of this technique is of great importance in the transmission of digital TV signals, because signal attenuation during data transmission must be avoided. The standard technique used is to modulate the signal of a carrier wave, according to the characteristics of the information to be transmitted, and the wave works with various frequency bands, reducing interference, distortion and attenuation, which are harmful to audio and video transmissions, as is the case with digital TV.[8]

The modulation techniques used for digital signals can be divided into:

> **COFDM** *(Coded Orthogonal Frequency Division Multiplexing)*: used by European and Japanese standard technologies, its main advantage is its robustness against noise from other TV signals transmitted over the air.
>
> **8-VSB** *(8 Level - Vetigal Side Band Modulation)*: used by the American standard, it has the characteristic of diffusing the signal transmitted by satellite or terrestrial, but it is less immune to noise than COFDM.

The multimedia data that will be transmitted and received by the user needs to be processed in order to be transported, i.e. it needs to be encoded, compressed and decoded for transmission and display, which is necessary to maintain the transfer speed. There are a number of standards for this, known as MPEG.

2.1.3 AUDIO AND VIDEO CODING AND COMPRESSION

Audio and video compression is an essential activity for the dissemination of digital media and is carried out in the encoding stage. In order to adopt a coding and decoding standard *(codec*[9]), the relationship between time and compression rate must be considered.

According to Montez (2005), a great example of this is the MPEG family of standards, which are used for the coding and compression of multimedia data, of which there are three, MPEG-1, MPEG-2 AND MPEG-4:

- **MPEG-1** - Used in encoding for the VHS video standard, it has very low image decoding quality of up to 1.5 Mbps;
- **MPEG-2** - Based on MPEG-1, but much more sophisticated and improved. It can encode videos in TV quality, from 4 to 9 Mbps, up to HDTV quality, between 15 and 100 Mbps;
- **MPEG-4** - This model is a standard for encoding audio and video for Internet applications with less than 64 Kbps, such as video conferencing. It allows commands to be sent to manipulate objects in the scene, such as changing the background image of a video.[9]

2.1.4 INTERACTIVITY CHANNEL AND FEEDBACK CHANNEL

Interactivity requires that the communication channel allows information to flow in both directions, i.e. from the transmitter to the receiver and vice versa.

Digital TV channels transmit information from broadcasters or retransmitters to their users in two ways: *Broadcast* - uses broadcasting channels to communicate with several users; *Unicast - communication* is done point-to-point, i.e. transmitted directly to a single individual.

In the return channel, the information is transmitted in the opposite direction, from the user to the broadcaster, generating interactivity.

2.1.5 MIDDLEWARE

Another important concept is *middleware.*

> Middleware is a general term, usually used for a piece of software code that acts as an agglutinator, or mediator, between two existing, independent programs. Its function is to make applications independent of the transmission system. It allows various application codes to work with different receiving devices (IRDs). By creating a virtual machine on the receiver, the application codes are compiled into the appropriate format for each operating system. In short, we can say that middleware makes it possible for a code to work with different types of receiving platforms (IRDs) or vice versa (PAES et al., 2005-A).

According to Paes et al. (2005-B), *Middleware* must provide basic services, both for the end

---

[8] Since 1997 there has been a public Embratel satellite transmitting digital signals to specific satellite dishes.

[9] *Codec stands* for Encoder/Decoder, a *hardware* or *software* device that encodes/decodes signals.

user and for signal operators, within Interactive Digital TV Systems, for example:

- For the end user / STB (Set Top Box / Converter Equipment)
  - the browser:
    - Open standards (HTML 4.0; Java Script 1.1);
    - Proprietary Standards.
  - Platform Management:
    - Software updates;
    - *Shut down* the converter.
- For signal operators
  - o Translations / Redesign of web content for TV;
  - o Proxy (content control);
  - o Security;
  - o Subscriber management.

Interactivity is one of the most important features that digital TV can provide, and this feature is realized by *middleware,* which interacts with software in well-known languages such as HTML and JAVA. This can create an environment very close to that of the web within digital TV systems, making it familiar for the user to interact.

There is a *middleware* standard for digital signal transmission systems, and in the case of Brazil, which uses the ISDB standard, it uses the ARIB (*Association of Radio Industries and Business*) standard, and its main characteristic is that audio, video and data are multiplexed, i.e. they travel over the same transmission infrastructure.

2.1.5 INTERACTIVITY

Interactivity is the ability of a system to transmit responses in real time according to the user's needs. This feature in digital TV can be compared to the union of the Internet and television.

> Nowadays almost everything is sold as interactive; from advertising to microwave ovens. There is a growing interactivity industry. The adjective interactive is used to describe any thing or object whose operation allows its user some level of participation, or exchange of actions (PALÁCIOS, 2000).

According to Montez (2005), digital signal transmission can be divided into different levels, as described below:

- **Level 0** is the stage at which the television displays black and white images and has one or two channels;
- In **Level 1,** television becomes colorful and the remote control appears (before contemporary web browsing), facilitating the viewer's control over the device;
- At **Level 2,** some peripheral equipment has been added to television, such as VCRs, portable cameras and electronic games. The viewer gains new technologies such as recording programs and watching or rewatching them whenever they want;
- **Level 3 shows** signs of interactivity with digital characteristics. The viewer can then interfere in the content via phone calls, faxes or e-mails.
- **Level 4** is the stage of so-called interactive television in which you can participate in the content from the network in real time, choosing camera angles, different routes for the information, etc.

These levels can be seen in Figure 2, which shows the applications related to the programs in relation to the interactivity channel.

	Program-related applications		
No interactivity channel	- Game score - Synopsis of novels, - Infotainfoes sobre jugadores	- E-commerce - Distance learning - Questions and answers	With interactivity channel

	Program-related applications		
	- Electronic programming guide - News and bulletins - Jc*gOS - Weather forecasting - Traffic	- Electronic connexion - Chat - Bank TV - E-commerce - NETWORKING	
	Applications not related to programs		

Figure 2: Examples of interactive applications[10]

According to the DTV (official SBTVD) website, Digital TV technology will enable features such as Mobility and Portability.

> **Mobility and portability** are characteristics that will put an end to the anguish of getting home quickly so as not to miss "That Show". Our Digital TV system allows programs to be viewed on buses, cars, boats, planes, laptops, on cell phones with viewers on the move, on desk tops in offices, or even with pocket receivers. (DTV, 2008)

The means of transmission of the Digital signal, via terrestrial, sends the TV signals over the air from stations in the television network (transmitter). A cell phone (cell phone) with a TV antenna and a Digital TV tuner (receiver) can pick up and translate these signals, as shown in Figure 3.

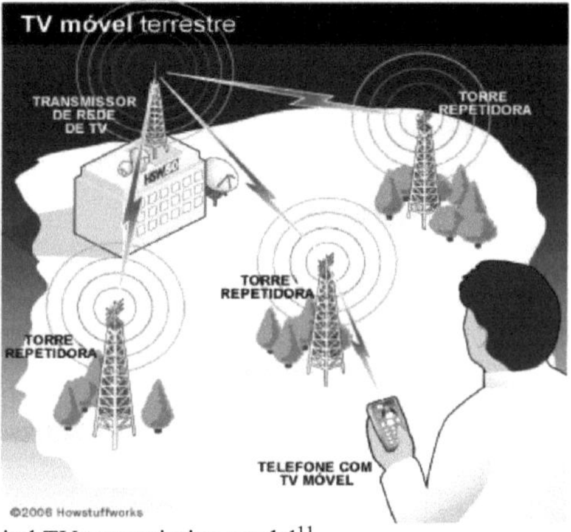

Figure 3 - Mobile Digital TV transmission model[11]

The DTV website (2008) also states that, in addition to the Mobility and Portability features, one of the highlights of digital transmission technology is Multiprogramming, as it allows activities such as: "[...] shopping on TV without having to use a telephone, voting in polls, consulting the broadcasters' program guide and other services that will appear as Digital TV is consolidated throughout the country".

Multiprogramming is an alternative to High Definition, which allows you to watch different programs on the same channel, or watch the same program from several different angles (very good for sports in general). To use this feature, you need a device for each program, because the audio is embedded in the program's video (DTV, 2008).

[10] Figure taken from Montez's book (2005)

[11] Figure taken from http://informatica.hsw.uol.com.br/telefones-com-tv2.htm.

Figure 4: Multiprogramming of the BBC Channel[12]

2.2 DIGITAL TV STANDARDS

This section describes the different standards that have already been implemented around the world, such as ISDB (*Integrated Services Digital Broadcasting*), adopted by Japan; ATSC (*Advanced Television Systems Committee)*, adopted by the USA, Canada, Mexico and South Korea and finally DVB (*Digital Video Broadcast), adopted by* the other countries that have already decided which standard to follow, especially those in Europe, Asia, Africa and Oceania.

According to Fundagao CPqD (2007), the number of households with analog TVs is very high, as can be seen in the table below. These figures can be considered a potential market for digital TV.

Table 1. Size of the analog TV market that could migrate to digital transmission.

Countries	TV households (millions)
USA, Canada, South Korea	125
Taiwan and Argentina.	15
European Union countries, Australia, New Zealand, Singapore and India	205
Japan	45
Brazil	38

2.2.1 ATSC *(ADVANCED TELEVISION SYSTEM COMMITTEE)* STANDARD

In the mid-1990s, the United States had its own transition process for digital terrestrial television, the ATSC, where it had its own selection process full of political pressures, industry pressures and criticism from advocates of technology that was not chosen, as in Brazil (Schoen 2006).

In the United States, the technology chosen was called eight-level vestigial sideband modulation or 8/VSB, this system is called ATSC.

ATSC is an open and published standard that anyone can implement, which means that other technologies can continue to exist in the ATSC world and that people can continue to think of new

12 Figure taken from www.bbc.co.uk/tv/.

applications that had not yet been imagined when it was originally standardized.

According to Pinheiro (2006), the advantage of having ATSC technology is due to its better and wider coverage where a higher transmission rate brings more data services. Lower costs for acquiring transmitters and lower operating costs, using existing transmission towers and allowing full use of the UHF and VHF bands.

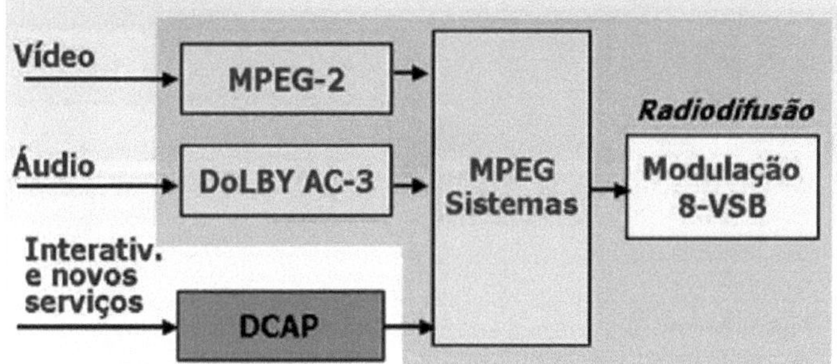

Figure 5 - Model of how the American digital transmission system (ATSC) works

According to the ADNEWS TV website (2008), the US standard promises mobility and portability, and will be demonstrated in Brazil in the near future:

According to ATSC, the transmission used in the standard offers the same coverage, including reception with an indoor antenna, with 40% of the transmission power of other systems; greater transmission capacity (19.4 Mbps), guaranteeing greater possibilities for transmitting new video content in HD or SD and also information services; lower transmitter acquisition costs and lower operating costs.

ATSC has an advantage in social terms because of the low cost of the set top boxes, averaging $50.00. The equipment would make it possible for a large part of the population to receive both high-resolution (HD) and standard-resolution (SD) signals, guaranteeing better image quality on analog TVs, interactivity and the availability of new content.

Moreira (2006) reports on the change of TV transmission standard in the USA (from analog to digital) for various reasons:

The ATSC standard implemented in Canada, South Korea, Mexico and mainly in the United States, has this technology that produces images in a 16:9 format, with up to 1920x1080 pixels *Wide Screen* with a resolution of up to 6 times more than an ordinary television with the NTSC standard.

The definition of this standard has the possibility of evolving the sound quality to match that of home theaters, with Dolby Digital systems transmitting up to 6 virtual channels in standard definition.

2.2.2 DVB (*DIGITAL VIDEO BROADCASTING)* STANDARD

According to DVB (2008), the DVB standard is the European standard most widely used by the main private TV operators. It was created to meet the diverse needs of various countries, and for this reason it is characterized by being a very flexible standard in terms of configuration modes, allowing the transmission of various video qualities in an 8 MHz band, from SDTV to HDTV, for which reason there are between 250 and 300 members from 35 countries dedicated to developing the DVB standard.

global digital TV delivery standards and associated services.

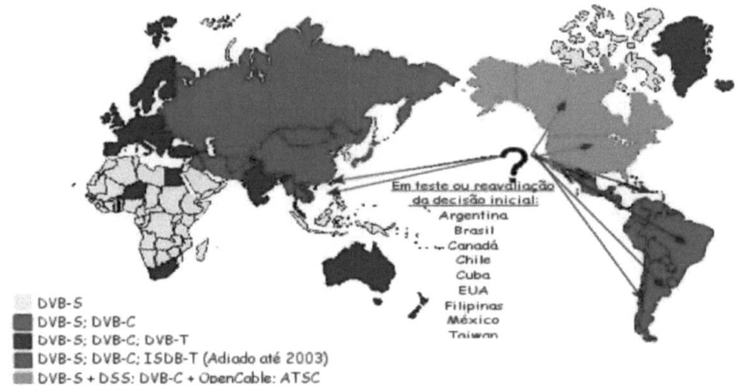

Figure 5 shows the geographical distribution of DVB around the world (taken from http://www.midiacom.uff.br/itvsoft/pdf/paes_2005a.pdf).

The DVB standard is transmitted by the following broadcasts:

- DVB-C (**cable** transmission);
- DVB-T (**Radio Frequency** Broadcasting);
- DVB-S (DVB-S2) (**Satellite** transmission);
- DVB-H (Standard that adapts DVB-T to bring DBV technology to portables);

According to Moreira (2006), it is known for being more versatile, making it easier to transmit multiple virtual channels on the same frequency. It operates at a frequency of 8 MHz, which puts it at a disadvantage compared to the Japanese and American frequencies, which operate at 6 MHz.

According to Taguspark (2006), DVB aims to introduce digital television broadcasting using cable (DVB-C), satellite (DVB-S) and terrestrial (DVB-T) transmission, providing features such as:

- Asymmetric, high-speed Internet access at a low cost;
- Transmit standard definition images (SDTV) on narrowband channels;
- Provide a technological evolution platform for interactive systems.

The resources that this standard makes available to users can be grouped into two lines of thought: "Multiprogramming" and "Interactivity", as shown in Figure 6.

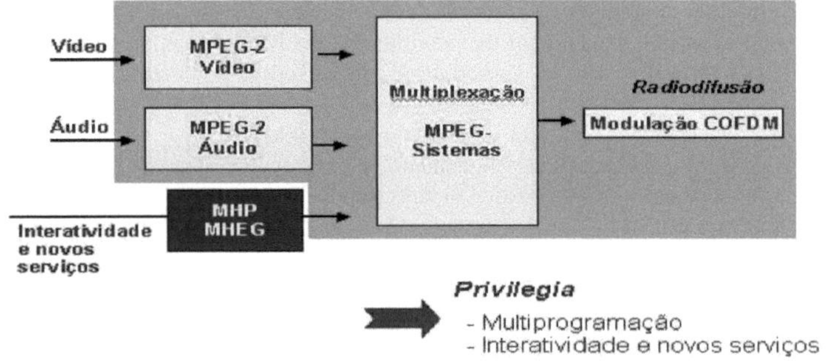

Figure 6: Model of how the European digital transmission system works[13] .

Taguspark (2006) also comments that in addition to the interactivity and multiprogramming resources, the technology is capable of:

Provide security mechanisms that can be used in services with sensitive and/or paid content..;
To meet today's interference protection requirements.

An important detail mentioned by Taguspark (2006) is that DVB uses the MPEG-2 standard for video and image compression. A very important component used by DVB. This is a process whereby the signal can be controlled and directed to certain groups of users.

> Before being transmitted, the video signal goes through a process of scrambling and encryption. In other words, at the source the video signal is encrypted and scrambled so that it is incomprehensible to a decoder who does not know the secret with which it was sent. There are two factors that make the content unreadable to anyone who doesn't have the secret: the "Control Word" (CW), which is used to decrypt the signal, and a key to decipher this CW (Taguspark, 2006).

2.2.3 ISDB (*INTEGRATED SERVICE DIGITAL BROADCASTING)* STANDARD

The Japanese government, together with its research body, was largely responsible for the development of ISDB-T technology, investing most of the money financially, with the help of the NHK TV channel.

Studies for the development and implementation of the model began in 1994 and were only finalized after approximately four years. Based on this prototype, a digital antenna was installed on the Tokyo transmission tower in 1999 for preliminary tests. After several field tests and studies on the acceptance of the standard to be implemented, the model was approved.

The creators of Japanese digital transmission technology (the Japanese government and broadcaster NHK) used the qualities of the previous systems, such as
DVB, for example, and have made major improvements, making the system complete and robust.

One of the differences between the Japanese standard and the others, according to Souza (2007), is that the system integrates all forms of broadcasting services into a common data channel, which can be transmitted by satellite, cable or terrestrial means. This standard can be divided into three different types of service: sound services, audio services and interactive data services. According to Oliveira and Ribeiro Jr. (2006), the great advantage of the Japanese system is its mobile and portable reception, i.e. it is possible to receive digital TV programming on mobile devices.

The website of Japanese broadcaster NHK[14] (2008) also states that the ISDB system's outstanding features are its ability to provide a variety of information, separating it into three levels: video, sound and data services, and being resistant to any interference encountered during portable or mobile reception.

> In transmission, the system must be intelligent enough to separate the information to be sent (sound, video and data), be flexible to accommodate different service configurations and ensure flexible use of transmission capacity, and be available for future needs to be met and compatible with analog and other digital services (NHK, 2008).

Sarmento and Reis (2007) comment that the most important point of the ISDB standard is its versatility, for example: data transmission can be received by a PDA or a cell phone, or even through the use of a computer or home server, making it possible to access television program websites and even *download* receiver update services.

The Medialess website (2008) defines the peculiarities of the ISDB system as the flexibility to change parts of the original project (such as the choice of audio and video codecs) and support for mobile systems, allowing cell phones, notebooks and handhelds with built-in ISDB-T receivers to tune in to TV channels on the move.

Figure 7 shows how the Japanese ISDB system transmits. The video, audio and data services are split up and then go through the *multiplexing* process (MPEG).

Japanese system: ISDB

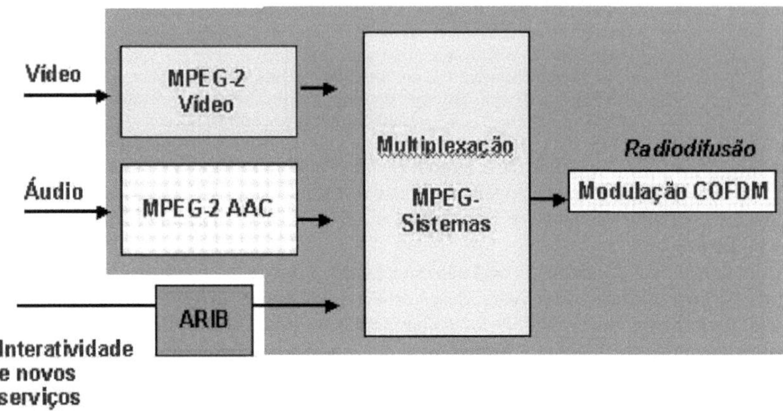

Figure 7: Working model of the Japanese digital transmission system[19] .

Because of these distinctive features, by mid-2004, 12 million Japanese homes already had digital TV sets or decoders. And sales are expected to be high by 2010, as this system has enjoyed great success. Currently, Japan has around 20 channels that are broadcasting using 100% of the Digital TV resources, including: NHK General, NHK Educational, Nippon TV, TV Asahi, TBS, TV Tokyo, Fuji TV and Tokyo MX TV.

[19]Taken from the ANATEL website, available at the following link: http://www.anatel.gov.br/Portal/exibirPortalInternet.do

Chapter 3

3. SBTVD - SIST. BRAZILIAN DIGITAL TELEVISION SYSTEM

SBTVD is the digital terrestrial signal transmission system used in our country, known in the market as ISDB-T (*Integrated Services Digital Broadcasting - Terrestrial*).

In 1999, Anatel (National Telecommunications Agency), together with CPqD, began the process of evaluating digital signal transmission technologies in the world, looking at technical and economic issues.

After several studies carried out at various universities, in 2005 the SBTVD was presented to the then Communications Minister Hélio Costa, which was developed based on ISDB-T (the system used in Japan).

But it was presented to the Brazilian people at the end of 2007, and was concentrated only in the city of Sao Paulo.

3.1. CHOOSING THE DIGITAL TV STANDARD

According to Cruz (2008), the Japanese ISDB-T standard is the newest and best performing technology. This finding had already been made in a set of tests carried out by Mackenzie University in 2000, comparing Japanese technology with American standards (ATSC) and European standards (DVB-T).

Cruz (2008) also states that American technology has not shown significant results in reception with an internal antenna. European technology, despite having points in common with Japanese technology, does not allow the digital signal to be transmitted to cell phones, requiring specific channels for mobile transmissions, which would not be feasible in our country. During the selection process, the researchers defended the Japanese standard as the best for Brazil, as it met all the requirements that were established.

3.2. IMPLEMENTATION PROCESS IN BRAZIL

The process of implementing digital TV in Brazil is recent and is being carried out as follows: **High Definition** (HDTV) with *widescreen* format and **Standard Definition** (SDTV) with traditional format.

For this reason, Cruz (2008) defines the debut of digital TV technology in Brazil as "disappointing", since the only feature presented to consumers was high definition in its display.

> Of the four main new features, Brazilian digital TV only offered high definition at the premiere. Commercial broadcasters were not very interested in multiprogramming, as it fragments the audience, which is not good for the advertising-based business model. The interactivity software developed in Brazil, called Ginga, was not adapted by television manufacturers in time for the launch. And there are no cell phones on the market that allow digital TV to be received in the Japanese format and that work on mobile phone networks. (Cruz, 2008)

According to Brain (2007), the plan to implement SBTVD began in 2006, but the implementation process will take approximately 10 years, since it will only be in 2016 that the analog system will cease to be transmitted by broadcasters and retransmitters in our country.

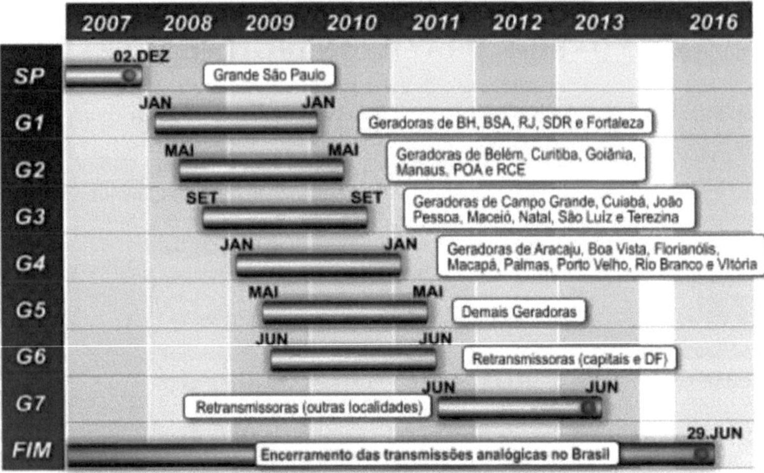

Figure 8. Brazil's digital TV deployment schedule (Brain, 2007)

One of the difficulties in implementing SBTVD on cell phones, according to Fuoco (2007), is the lack of encouragement from telephone operators, because with the development of this technology, users would not make calls to watch their favorite program.

> The Minister of Communications, Helio Costa, has criticized the country's cell phone operators because, in his opinion, it is due to their lack of commitment that Brazil still doesn't have cell phone models with TV reception on sale. "I haven't seen any interest from telephone companies in charging for handsets," said the minister at a press conference. (Fuoco, 2007)

According to research carried out by FUNDAQÁO CPqD (2007), Brazilians have a high rate of access to analog TV and the southeast region has the highest percentage:

Table 2. Analog TV access rate (by region)[16]

Region	Households with TV (%)
South East	96
Center West	92
South	94
North East	88
North	88

Based on this information, the UOL website (2007) states that television is the most widespread means of communication in our country, reaching approximately 92% of the population.

> Television is one of the most popular consumer electronics in Brazil. Approximately 92% of Brazilian households have at least one TV set.[20]

In order to achieve a high level of acceptance in the market, one of the resources that the

[20] Taken from the CPqD Foundation website (2007)

government will adopt to encourage the deployment of SBTVD transmission resources is the exemption and financing of converters, in order to reach the needy population.

> "There is the possibility of selling [converters] for a monthly fee of between seven and ten reais," said Communications Minister Hélio Costa. "We're going to be able to put this funding in place for the explosion of digital TV in our country" (IDG NOW, 2007).

As one of the ways of promoting the digital signal and boosting sales of televisions with the integrated technology, the government is promoting advertising campaigns associating our country's new television signal transmission system with our national flag, as shown in the figure below.

Figure 9. Advertising campaign images[21]

In contrast to the Brazilian government's efforts to implement this technology, there is a large stock of analog television sets with a population that has low purchasing power . One of the solutions found was the construction of the "Set top box" - Figure 9, which is nothing more than a tool with social purposes , to offset the cost of the technological evolution we are experiencing.

3.3. BEHAVIORAL IMPACTS

Brazil's Digital Television System should contribute to the convergence of communications services, enabling the emergence of services resulting from digital technology and technological updating, as well as opening up the job market for professionals who are prepared to work with this technology (MOTA and TAKASHI, 2005).

Cruz (2008) states that with the development of the SBTVD, a job market will emerge that can be exploited by various professionals in our country.

> Brazilian software companies have the potential to export interactive applications to Japan. For Yasutoshi Miyoshi, from Primotech21, there would be an opportunity, especially if the application was linked to a theme identified with Brazil, such as soccer. "The Japanese would want to know how the soccer country interacts with television during matches."

Moreira (2007) says that advertising and publicity on digital TV have also undergone some transformations and will have a major impact.

behavior in the Brazilian population, because together with the commercial

that we have visualized , the television networks will be able to send to the viewers some additional information about the advertised product or the advertiser, contact telephone numbers, store addresses, among other relevant data for the potential consumer.

> If the user likes the outfit that an actress is wearing in the soap opera, it will be possible, at least technologically, to buy that outfit in one click. "An icon could appear in the corner of

[21] Taken from http://www.vivasemfio.com/

the screen, indicating that the product is for sale. Those who want to know more click, those who don't continue watching the soap opera uninterrupted," explains Rodrigo Araújo, head of EITV, a software developer for digital TV [...] (Moreira, 2007).

These resources can be seen in the figure below, which shows a movie about obesity and allows viewers to see advertisements on this subject, such as: Spas, Diet Tips, Gyms...

Figure 10. Image from American broadcaster Fox[18]

According to the DTV website (2008), digital TV technology will have major behavioral and social impacts in our country:

> Economically, socially, scientifically, politically, technologically, commercially and any other adverb ending in "MIND" will be affected by this new way of watching TV (DTV, 2007).

Moreira (2007) comments that broadcasters are also investing in make-up, sets and special equipment to avoid a "reality shock" in programs recorded in HD.

> Make-up artist Guilherme Pereira, who is working on retouching the actors in Rede Bandeirantes' soap opera "Dance, Dance, Dance", recorded in high definition, says that "you have to be very careful. The products are special for HD, imported from Germany", reveals the specialist, who has brought the experience he acquired in the cinema to the screen (Moreira, 2007).[22]

3.4. SBTVD AND THE CHALLENGE OF DIGITAL INCLUSION

Cruz (2007) believes that digital television creates an opportunity to bring interactive services to a portion of the population that currently does not have access to a computer. In his report, he states:

> When the return channel is regulated, television will theoretically work like a computer [...] in Japan, all the banking services available on the Internet can also be accessed by cell phone and television (Cruz, 2007).

Batista (2005) states that with the development and implementation of the system SBTVD, we will be able to reduce (digital) social exclusion:

> [...] the low-income population will be able to benefit enormously from the most advanced digital television technologies, such as large high-definition monitors, portable TV sets, mobile TV sets and TVs with a high level of interactivity and Internet connection, through collective use in schools, libraries, community centers, trade unions, government service stations, transport equipment, etc. (Batista, 2005) [...]

[22] Image taken from: http://www.fox.com.

On the other hand, Mota (2005) does not believe that the SBTVD system will promote social (technological) equality among the poor:

> "Social inequality in the field of communications, in the modern society of mass consumption, is not only expressed in access to material goods - radio, telephone, television, Internet - but also in the user's ability to make the most of the potential offered by each communication and information tool, based on their intellectual and professional skills (Mota, 2005)."

3.5. ADVANTAGES AND DISADVANTAGES OF SBTVD

Cruz (2007) believes that one of the advantages of digital TV technology is interactive transactions: "[...] Internet banking is already reaching stagnation," he said. The research center is working on interactive digital TV applications.

> SBTVD will offer Brazilians several advantages over analog television, such as better sound quality and image clarity, optimized coverage, greater programming diversity due to the greater number of channels available, Internet access and, above all, a new range of interactive applications encompassing text, video, audio and the most varied types of graphic elements (Souza and Oliveira, 2005).

According to Moreira (2007), Digital TV technology will have a major impact on our country's television stations, as the cost of maintenance and transmission is very high and consequently broadcasters will not be changing their equipment quickly enough to meet this technology's needs.

> Producing high-definition commercials is also a possibility, but it involves costs. "Production companies and agencies will have to invest in equipment," points out Alvarez, from ABP. The broadcasting of ads in high definition and with interactive content will also have to be priced differently, according to José Marcelo Amaral, vice-coordinator of the SBTVD Forum's market area and Record's technology director (Moreira, 2007).

Moreira (2007) conducted an interview with sports journalist Chico Lang, who jokingly mentions a disadvantage of implementing SBTVD, because with high definition, wrinkles will appear on TV. "Nobody over 40 is going to get a job on TV," says the presenter. "This is already a trend, which will only get worse with high definition," he adds.

Chapter 4

4. INTERVIEWS WITH PROFESSIONALS

This chapter presents the results of a series of interviews carried out in 2008 by our team, with professionals working in the market who are witnessing the changes and developments that Brazilian television is experiencing, in order to draw a parallel between theory and market practice. The result is a better understanding of this technology in a practical way, verifying the changes that are being made to adapt to this new system.

Of all the professionals consulted and sought out for the development of this work, we have selected just a few interviews that demonstrate the proposed theme of the work in a practical way.

- ALTAIR CARLOS PIMPÁO - owner and General Director of TV GALEGA, a cable television station in the city of Blumenau - Santa Catarina;
- DENIS HIPÓLITO DE ARAUJO - Project supervisor at Positivo Informática S/A, working in the R&D Project Management area, coordinating projects in the Digital TV area;
- EDUARDO BICUDO - was once director of engineering at TV Globo and is currently vice-director of education at SBET (Brazilian Society of Television Engineering) and a member of the Digital TV forum;
- MARCUS MANHÁES - has been participating in research and development projects in wireless telecommunications at the CPqD Foundation, including digital radio systems, cellular telephony and digital television;
- NANI MORENO - is Marketing Coordinator at Rede Massa - TV Naipi (SBT affiliate).
- RICARDO MUSSE - is a professor in the sociology department at the University of Sao Paulo. He holds a doctorate in philosophy from the University of Sao Paulo (1998) and a master's degree in philosophy from the Federal University of Rio Grande do Sul. He is currently following the new market trends resulting from Digital TV.

4.1 ALTAIR CARLOS PIMPAO[19]

Journalist Altair Carlos Pimpao is the owner and Managing Director of TV GALEGA[20] , which is part of the cable television system in the city of Blumenau - Santa Catarina. This venture represents an advantageous space for publicizing companies, products and services, given the significant number of viewers it reaches (138,000/Feb-00), just locally, not counting audiences from other locations and Internet users, according to the station's website.

I - What are the image differences between analog TV and digital TV?

In analog TV, images are transmitted with a resolution of 480 horizontal lines. In digital or high definition TV there are up to 1,080 horizontal lines. The screen of digital monitors changes from 4:3 to 16:9.[23]

II - What do you see as the future of Brazilian television with the technology that is currently being implemented?

By 2016, all free-to-air broadcasters will have to switch to the new system. In addition to better image quality, which will be clear, we will no longer have transmission noise and high definition, and we will be able to watch at a distance of 3 times the height of the screen, whereas today the recommended distance is 7 times the height of the screen. Interactivity will be another breakthrough. The viewer will be able to choose programs, for example. The last time I was in Germany, I saw my daughter checking - on the television set - whether her husband's flight was on time. I was also able to find out the name of a song in a movie, the soccer results and details of some programs.

It's already known that from 2010 it will be possible to pay bank bills, shop online and contact the government. Viewers will be able to vote for a freshman, eliminate Big Brother, schedule time to watch their favorite team play and buy a product used in a soap opera. Being able to watch television on a cell phone is the other great advantage of digital TV.

[23] Contact e-mail: Altair.Pimpao@tvgalega.com.br

III - How does this affect the creation of scenarios?

Television stations are increasingly turning to virtual sets. The cheapness of the programs and the boards needed for them will certainly lead to the use of virtual sets only for programs made in the studio. This will affect the lives of set designers, carpenters, painters, etc.

IV - Will Digital Television bring more or less costs for generating scenarios?

From my previous answer, you can already see that I believe it will cost less, once the equipment, which is already around R$20,000.00, with character generator and other technical resources, has been paid for.

V - With the Digital TV image, what are your concerns with make-up, costumes and set design?

I don't know yet. Those who produce soap operas will be able to say for sure. But I think that because of the better quality in reproducing people's natural colors, it won't even be necessary to go too far with the make-up. Costumes will continue to be important and it's what makes Rede Globo's soap operas stand out.

VI - What challenges and difficulties does the migration from analog to digital pose for the country's broadcasters?

The initial difficulty is the cost of the equipment that broadcasters will have to buy. I've heard that BRDE is going to launch a line of credit at cheap interest rates to make this easier. Then comes the question of televisions. Our people's purchasing power doesn't allow anyone to buy a 46" Sony for 6,000 reais.

But it's likely that Brazilians will go to the credit union and pay 12,000, as long as it's on time. Those who subscribe to cable TV still have to buy the decoder. The information I have is that NET's costs R$799.00 and TVA's R$399.00, payable in 10 installments.[24]

VII - With the implementation of digital TV, will it be possible to develop software that interacts with the viewer?

From what is already known about interactivity, there is already software for it.

VIII - What are the likely characteristics of this new business model that is emerging for TV stations?

As the owner of a local closed-signal station, practically just a television programmer, our HD broadcast is still a long way off. That's why I haven't started dreaming about the possibility of revenue. I believe that *pay-per-view is* going to grow a lot, that broadcasters will be able to make money from the sale of products that are shown in soap operas or other programs, from *merchandising that can* be seen and read, even at a tiny size, and from participating in the telephone company's income from calls to make guesses or take part in draws, something that is already happening.

IX - At this point you can make your final remarks and thanks.

I hope I have contributed, albeit modestly, to the success that I am sure will come from your work, which is very well grounded in what I have read and learned.

4.2 DENIS HIPÓLITO DE ARAUJO[21]

He has a degree in Electrical Engineering from the Federal University of Paraíba (2002) and a master's degree in Computer Engineering from the Federal University of Paraíba (2003). He is currently a Project Supervisor at Positivo Informática S/A, working in the R&D Project Management area, coordinating projects in the Digital TV area. He has experience in Electrical Engineering, with an emphasis on Real-Time Systems, Embedded Systems and Distributed Systems.[25]

I - What are the image differences between analog TV and digital TV?

The signal transmitted is digital, so we have a signal (100%) or not. Unlike analog broadcasting, where the image showed flickers and ghosting.

II - What do you see as the future of Brazilian television with the technology that is currently being implemented?

With digital TV we will have better image and sound quality. However, this appeal is not enough for us to have mass adoption of the new technology. With the incorporation of the Ginga *Middleware*, we could see faster adoption of the technology by users. However, as with any new

[24] The broadcaster's website is available at the following link: www.tvgalega.com.br

[25] The curriculum is available on the web at: http://lattes.cnpq.br/1588157250582074.

technology, adoption is not immediate and it will take some time for there to be a significant number of new users for digital TV in Brazil.

In the future, advertisements will be divided into NCL and Java applications and traditional forms of advertising. We will also have the possibility of interacting with TV programming via any device (STB, cell phone, etc.).

III - What major impacts will digital TV technology have on advertising companies?

Advertising companies need to learn how to use this new means of promoting products, where users can interact more easily and give their opinions on programming.

There will also be the generation of new content by broadcasters, with the aim of retaining the user's attention in the programming. And also to combat the dissemination of video and image programming over Internet networks.

IV - Will Digital Television bring more or less costs for generating scenarios?

Initially, we will have to prepare the mentality of organizations for this new environment, and train people who are qualified to create content and new scenarios for this new digital paradigm.

V - With the Digital TV image, what are your concerns with make-up, costumes and set design?

As we have a better definition of the image, the concern with costumes and make-up increases, as the smallest details will be more easily perceived than with analog transmissions.

VI - What challenges and difficulties does the migration from analog to digital pose for the country's broadcasters?

The biggest difficulty is winning over the audience in a short space of time. There is a lack of policies to explain to users the benefits of the new media, and to try to attract people to migrate to the new system.

VII - With the implementation of digital TV, will it be possible to develop software that interacts with the viewer?

The Ginga *Middleware* will be the tool to fulfill this objective. However, Ginga has problems with rights to use the part intended for Java applications. Sun, in partnership with the SBTVD, is developing a proposal for a Ginga-J free of copyright payments. Until this becomes a reality, there are discussions about releasing a Ginga version (Ginga Ready or Ginga 1.0) with only Ginga-NCL and Lua support.

VIII - What are the likely characteristics of this new business model that is emerging for TV stations?

The possibility of increasing the number of ads with interactive applications, incorporating new user information, usability tips, navigability, games, online shopping, etc.

4.3 EDUARDO BICUDO

Eduardo Bicudo was once director of engineering at TV Globo and is currently vice-director of education at SET (Brazilian Society of Television Engineering) and a member of the Digital TV forum. He was one of the people responsible for the tests carried out by ABERT/SET in conjunction with the *MACKENZIE* faculty to define the Digital TV Transmission System in Brazil.

I - What is the difference between the Japanese standard and the Brazilian standard?

Our standard has changes (several evolutions). If a Brazilian consumer buys a Japanese television set, it won't work here in Brazil, but if a Japanese consumer buys a Brazilian television set, it will work perfectly in their country. It's the same problem that exists with program versions (...), the Japanese standard version is 0 and the Brazilian version is 1. The Brazilian version works perfectly in Japan, but the Japanese version doesn't work here in Brazil.

II - Why is the digital signal deficient and not working in certain places in the city of Sao Paulo?

We can compare this technology to our cell phones. In certain places in the city, the cell phone doesn't receive a signal and is out of the area, and the same thing happens with digital transmission. If there is any interference in the middle of the transmission, there will be no communication and the television won't work [...], in a few years we won't have this kind of problem.

III - Why are many consumers currently experiencing problems with the installation of the digital TV system and the set-top-boxes?

You could say that Brazilian Digital TV is a child. It was born and today it is learning to walk.

SBTVD is in a learning phase and in the near future we won't have problems similar to the ones we're seeing.

IV - Will we be able to watch all the channels that are currently on pay-TV on digital TV?

With digital broadcasting, the rules remain the same. The rules established by the Ministry of Communications require media companies to obtain an authorization, which we call a concession. The concession is divided into two levels as we already know: open TV (which includes channels such as GLOBO, SBT, RECORD and BAND) and pay TV (which includes channels such as SPORT TV, ESPN, DISCOVERY CHANNEL). In digital broadcasting, only free-to-air TV channels are allowed to be shown.

V - What's wrong with using the "Collective Antenna" to deploy digital TV in buildings and condominiums?

For the time being, there are no rules for installation. Each antenna installer installs the digital system in condominiums, buildings or associations, according to their own convenience, in order to serve consumers with a precarious installation. We can compare this problem with what we saw in the 1960s when we only had the VHF transmission system. With the advent of UHF broadcasting in the 1980s, consumers' televisions could not be tuned to receive this signal. In an attempt to get around this situation, antenna operators installed converters in buildings and/or condominiums that received the UHF signal for VHF. Unfortunately, many antennas will be doing "this workaround" by converting the digital signal into analog, in order to serve customers cheaply.

VI - Briefly, what are the stages of digital transmission in the case of collective antennas?

We can define three extremely important stages for the transmission of Digital TV in condominiums and buildings: Capture, Processing and Distribution.

4.4 MARCUS MANHÀES22

Marcus Manhaes has worked at CPqD since 1984; he obtained a master's degree in education from the State University of Campinas (UNICAMP) in 2003. At the CPqD Foundation, he has been involved in research and development in *wireless* telecommunications projects, including digital radio systems, cellular telephony and digital television. He has carried out numerous international certification projects on telecommunications equipment, as well as developing methodologies for interoperability testing, benchmarking and optimization in *CDMA* and *GSM* cellular telecommunications networks. He is currently involved in the development of the Interactivity Channel System for Brazilian Digital Television.[26]

I - What are the image differences between analog TV and digital TV?

Analog TV can display signals with noise and interference in the image. The resolution of the image and the ability to show detail are only reasonable, but viewers are used to this quality because of the system's long existence.

Digital TV brings new features in the quality of signal reception, which is now more robust, without "flickers" and "ghosts", examples of phenomena that people were used to in analog. At the limit of poor reception quality, digital TV may simply show no signal at all, but it is able to operate perfectly with much weaker or disturbed signals.

The ability to express higher resolution images on digital TV has the greatest impact on the viewer. There are now receivers with large screens that practically bring the cinema into the home. The richness of detail in high-resolution digital images, known as Full HD, is the great attraction of digital TV, although robustness is a valued technical feature.

II - What do you see as the future of Brazilian television with the technology that is currently being implemented?

For broadcasters, signing up to the new digital terrestrial TV system means an immediate transformation in the way content is organized and broadcast. As soon as broadcasting in digital format begins, quality and diversity of content become crucial requirements. In this sense, the focus is still on content, an argument for which there is already a great deal of experience in Brazil, in the big networks. However, the future presents new possibilities for services based on the same television

---

[26] Researcher's email: (manhaes@cpqd.com.br)

platform. It is these services, still glimpsed in the minds of technicians, that will bring about the great transformation in so-called interactive TV. The characteristics of Brazilian society provide scope for e-government and social inclusion services to be strongly exploited in this innovative vehicle for communication and interaction that emerges with the digital TV platform.

III - How does this affect the creation of scenarios?

The preoccupation with scenery derives from high-definition broadcasting, where the high image resolution allows details to be visualized. Thus, those imperfections in the set, such as walls with retouching and seams, can be visualized by the viewer. However, this reasoning only applies to studio composites, where the total image presented will include the environment.

It should be noted that a well-established technique called *chroma key composites* images of people and objects with backgrounds generated by computer graphics, photographs or remote scenes. Image processing and complementary techniques should be considered as a strategy for composing scenes, and are increasingly accessible with the evolution of information technology and software for these purposes. Digitization therefore has an impact on the forms and procedures applied to content production.

IV - Will Digital Television bring more or less costs for generating scenarios?

The costs will be derived from the presentation option. When computer graphics and image compositing techniques can be applied, this will simplify set design immensely. However, by bringing the focus to details of the environment, this will certainly require more effort and, consequently, more expenditure on the execution and preservation of scenarios.

V - With the Digital TV image, what are your concerns with make-up, costumes and set design?

Again, high resolution is taken as the attribute that brings evidence. A practical way of understanding what comes with such an attribute is to admit that high resolution aims to reproduce reality, as if objects and people were physically in front of our eyes. So you know that in a face-to-face situation, you can see details of skin, clothing and everything else. You can even see the twist in the clown's mask. These details will also be observable on the high-definition screen. Naturally, professionals in the relevant fields will have to sophisticate their techniques in order to highlight or hide the details that interest them. This can mean new attributes to be exploited artistically, and such skills will make good professionals stand out.

VI - What challenges and difficulties does the migration from analog to digital pose for the country's broadcasters?

The cost of a digital TV transmission platform is the first obstacle for broadcasters. Therefore, financial support for small and new broadcasters is the first problem to be solved. The details and imperfections in sets, which require a large financial investment to overcome, only apply to the generation of programs and content in high definition. Small broadcasters can produce in standard definition until they manage to develop techniques that support them in producing in high definition. This brings us to another important challenge: professional training. It will be professionals who will build solutions, services and alternatives on the new television platform. Finding ways to develop and support technology-savvy professionals, in their specialties and fields, is the great challenge.

VII - With the implementation of digital TV, will it be possible to develop software that interacts with the viewer?

Perfectly. Experimental prototypes of these devices already exist. CPqD has developed experiments with banking and education applications. In addition to these, CPqD has developed innovative concepts that focus on usability, especially with regard to the inclusion of the physically and functionally disabled - in light of the potential of interactive digital TV - in the technology-society binomial.

VIII - What are the likely characteristics of this new business model that is emerging for TV stations?

In today's society, the TV is already a significant icon. With the new technological possibilities, the business model will have to be progressive and add even more value to the device. As is already evident, it is starting with digital broadcasting, with programming corresponding to what is seen on analog TV. Little by little, a transformation is taking place in the way content is produced, increasingly internalizing the productions of the film industry and gradually increasing interactive applications and services. However, we have to wait for the effects of the massification of digital receivers, full of their potential. It is likely that there will be changes in the number and characteristics of the players in the current value chain, including the emergence of players who centralize and distribute third-party content.

4.5 NANI MORENO

Nani Moreno is Marketing Coordinator at Rede Massa - TV Naipi (SBT affiliate).

I - What do you see as the future of Brazilian television with the technology that is currently being implemented?

TV will be even more versatile and with higher quality, interaction with the viewer will be much greater, in addition to the increase in audience due to portability, since it can be received by portable devices anywhere. TV, which is already a strong vehicle, is likely to become even more important and qualified, not least because it will be more segmented.

II - How does this affect the creation of scenarios?

A lot. With the format of the image in the digital system (16:9) compared to the current one (4:3), the aperture is wider and the quality of the image is much more specific. Broadcasters will have to adapt to the scenery, make-up and audio, which experts in the field have been studying for some years so that when producing a program the viewer doesn't identify flaws in the scenery, the image of the artist, etc.

III - Will Digital Television bring more or less costs for generating scenarios?

The scenery will have to be specific, especially the lighting. They are still carrying out studies on this, and over time the scenery will improve as they discover other ways of guaranteeing quality, and the higher the quality, the greater the investment.

IV - What challenges and difficulties does the migration from analog to digital pose for the country's broadcasters?

The adaptation of the antenna, which costs millions, engineering equipment as a whole (cameras, lighting, etc.), sets, editing, adapting the audience to this new concept will have to be done in the long term, training and adapting TV professionals to the new working method, among other things.

V - With the implementation of digital TV, will it be possible to develop software that interacts with the viewer?

Interaction with the viewer is one of the main reasons for digital TV, in addition to sound and image quality. But it's still in the testing phase because each receiving device (conventional TV, portable, cell phone, etc.) will have its own system for viewers to participate and buy advertised products.

VI - What are the likely characteristics of this new business model that is emerging for TV stations?

With interactivity, viewers will be able to receive more content in addition to the information broadcast, for example, seeing a summary of a soap opera or buying tickets for their favorite soccer match. This will enable a huge range of businesses and advertisements, because whatever a viewer likes to see on a presenter, for example, she can click on the remote control and request a purchase.

4.6 RICARDO MUSSE

He is a professor in the sociology department at the University of São Paulo. He holds a PhD in philosophy from the University of São Paulo (1998) and a master's degree in philosophy from the Federal University of Rio Grande do Sul (1992). He has experience in research and teaching in the areas of sociology and philosophy, with an emphasis on sociological theory, working mainly on the following topics: social theory, historical sociology of Marxism, critical theory of society, sociology and German philosophy.

I - Will free-to-air TV networks be the biggest beneficiaries of adopting the Japanese standard? For what reasons?

In principle, the Japanese standard tends to favor open TV networks, as it adapts better to the model of broadcasting via shared towers. In addition, their choice of the "high definition" format tends to inhibit content producers without resources or even businesses based on "small investments".

II - Will the decision in favor of the TV networks harm the diversification of content production that was expected with the implementation of the digital system?

In principle, no. The diversification of content depends above all on the distribution of the frequency spectrum. The "high definition" format tends to reduce the spectrum, but in any case, the redistribution of the spectrum is a political decision, which I think will only be defined in the next few years.

I don't see how the reshaping of TV through its digitalization can ignore, for example, society's demand for educational and community TV. I think there will also be significant commercial investment. The trend, even in the long term, is towards greater democratization of Brazilian TV.

III - At the same time, how will the decision affect the electronics industries and telephone companies, which advocated the implementation of the European standard?

Unfortunately, despite what the newspapers say, the Brazilian electronics industry doesn't exist. We only have imported assemblers, even when these "assemblers" are national, as is the case with Gradiente. The implementation of digital TV is a great opportunity for us to create semiconductor factories and start producing the components of the electronics industry here.

The telephone companies, in principle, still don't have what they don't have - a decisive role in the distribution of images for TV. The great advantage of the open TV system is the sharing of transmission, a system set up during the military regime in the name of national integration, by the then state-owned Embratel.

We get rid of the risk of having to pay for the TV signal the absurd amounts we pay for using privatized telephony. But that doesn't mean that telephone companies don't have a place in the system. In digital TV, the return channel is essential, and this, due to the difficulties in "cabling" a country the size of Brazil, will mostly be the telephone line.

IV - What assessment can be made of the counterparts offered by Japan, such as the installation of electronic component factories and the transfer of technology?

I can't say much about this because the data has not yet been released in full.

V - What will be done with all the technology developed in Brazil for digital TV, especially the adapters for conventional sets? Will it be adapted to the Japanese standard or has it become a useless investment? Has the idea of Latin America having a standard developed on its own continent become unfeasible?

The Brazilian technology developed will be used in the production of decoders (or converters). The income profile of the Brazilian population tends to demand its own solutions, i.e. creative and less expensive ones. In addition, it seems that opting for the Japanese standard leaves room for the use of national software, in what experts call "middleware".

VI - What will the transition phase be like for broadcasters? And for the viewing public? Will it be a slow adaptation, as has already been estimated?

The transition to digital TV will be slow all over the world. For two reasons: (a) the economic cost of replacing production, transmission and reception equipment; (b) the acceleration of technological innovation in recent years. We still don't know, for example, what kind of convergence we will have in the future. Will the same piece of equipment work as a way of accessing the internet, as a substitute for a landline telephone and as an image receiver?

Is that what the consumer wants? These are decisions that will only be consolidated in the long term.

VII - Can the government back down from its decision due to pressure from organized sectors of society, in a reaction similar to the one that postponed the start of work on the transposition of the San Francisco River?

This, like any democratic government, is subject to pressure from organized society and public opinion. I think, however, that the discussion in the country has not yet acquired enough strength to mobilize in favour of a particular option. Furthermore, I think it would be more appropriate to unite the forces of organized society on the issue of the "business model", after all our main objective should be the democratization of the media.

4.7 CONCLUSION OF THE INTERVIEWS

The purpose of the interviews was to identify and interpret the possible reasons why Japanese technology was chosen to be implemented in our country. Each of the interviewees mentioned improvements and difficulties that we will face in this new phase in which we are living with the changes to our television signal transmission system.

Altair Carlos Pimpao, director and owner of a TV network, pointed out that the technology being implemented in our country will bring portability (the possibility of watching television on a cell phone) and interactivity.

Pimpao commented that the market expects that from 2010 onwards it will be possible to pay bank bills, buy online and contact the government. Viewers will be able to vote for a freshman, eliminate Big Brother, schedule time to watch a soccer match, or even buy a product used in a soap opera.

Nani Moreno, marketing coordinator for Rede Massa (an SBT affiliate), believes that the digital signal will not be implemented immediately in broadcast networks outside the major centers, because the adaptation of the antenna, the purchase of engineering equipment as a whole (cameras, lighting, etc.), the construction of new sets, the editing of programs, have a high cost, making it unfeasible to acquire this equipment in this period.

Denis Hipólito de Araujo believes that digital TV will only become interactive with the incorporation of the Ginga *Middleware*, as we will be able to see a faster adoption of the technology by users. However, as with any new technology, adoption is not immediate and it will take some time before we have a significant number of new users for digital TV in Brazil.

One of the difficulties pointed out by Eduardo Bicudo is the installation of the digital system in condominiums, buildings or associations, as there are no installation rules.

Marcus Manhaes points out that the cost of a digital TV transmission platform is the first obstacle for broadcasters. Small broadcasters can produce in standard definition until they manage to develop techniques that support them in high definition production. This brings us to another important challenge: professional training. It will be professionals who will build solutions, services and alternatives on the new television platform. Finding ways to develop and support technology-savvy professionals, in their specialties and fields, is the great challenge.

Ricardo Musse defends the Japanese standard as the best system on the market, because this technology is better suited to the model of broadcasting via shared towers. In addition, their choice of the "high definition" format tends to inhibit content producers without resources or even businesses based on "small investments".

Musse believes that Brazilian technology will be applied in the production of decoders (or converters), because the income profile of the Brazilian population tends to demand its own solutions, that is, creative and less expensive ones.

Table 3 shows a comparison of the interviews conducted with the professionals.

Table 3. Comparison of interviews

QUESTION	ALTAIR CARLOS PIMPÀO
1	High definition

2	Interactivity
3	The lives of set designers, carpenters and painters
4	It will cost less once the equipment has been purchased
5	You don't know
6	The cost of equipment
7	There is software for interaction.
8	HD broadcasting is still a long way off
	DENIS HIPÓLITO DE ARAUJO

1	The image does not show flashes or ghosts
2	Advertisements being divided into NCL and Java applications, as well as interactivity
3	Advertising companies need to acquire the knowledge of how to use this new medium
4	Initially, we will have to prepare the mentality of organizations for this new environment, and train qualified people
5	Small details will be more easily perceived than with analog transmissions
6	The biggest difficulty is gaining an audience in a short space of time
7	The Ginga Middleware will be the tool to fulfill this objective
8	The possibility of increasing the number of ads with interactive applications

	MARCUS MANHÀES
1	Richness of detail in high-resolution digital images
2	New possibilities for services based on the same television platform.
3	Image processing and complementary techniques should be considered as a scene composition strategy
4	This can mean new attributes to be explored artistically, and such an aptitude will set the good professionals apart.
5	The cost of a digital TV transmission platform is the first obstacle for broadcasters.
6	Experimental prototypes of these devices already exist. CPqD has developed experiments with banking and educational applications.
7	In today's society, TV is already a significant icon.

	NANI MORENO
1	TV will be even more versatile and quality, the interaction with the viewer will be much greater, in addition to the increase in audience due to portability
2	With the image format in the digital system (16:9) compared to the current one (4:3), the gap in visibility is wider and the image quality is much more specific, and broadcasters will have to adapt.
3	The backdrops will have to be specific, especially the lighting, which is still being studied.
4	The adaptation of the antenna, which costs millions, engineering equipment as a whole (cameras, lighting, etc.), scenery, buildings, adapting the public to this new concept will have to be done over the long term
5	Interaction with the viewer is one of the main reasons for digital TV, apart from sound and image quality.
6	With interactivity, viewers will be able to receive more content in addition to the information broadcast, for example, watching a summary of a soap opera or buying tickets for their favorite soccer match.

Questions

1 - What is the difference between analog and digital TV?

2 - In your view, what is the future of Brazilian television with the technology that is currently being implemented?
3 - How does this affect the creation of scenarios?
4 - Will Digital Television bring more or less costs for generating scenarios?
5 - With the image of digital TV, what are your concerns with make-up, costumes and set design?
6 - What challenges and difficulties does the migration from analog to digital pose for the country's broadcasters?
7 - With the introduction of digital TV, will it be possible to develop software that interacts with the viewer?
8 - What are the likely characteristics of this new business model that is emerging for TV broadcasters?

Chapter 5

5. EXAMPLE OF AN INTERACTIVE APPLICATION

At this stage, an application was developed that simulates the interactivity resources proposed for the implementation of Brazilian Interactive Digital TV over the next few years, according to the timetable and deadlines set by the Ministry of Communications.

The two main broadcasters in our country, TV Globo and SBT - Sistema Brasileiro de Televisâo, were taken as an example for the development of this application, in view of their local programming and possible interactivity features.

5.1 CHOICE OF TOOL AND SYSTEM STRUCTURE

The structure of the system follows the line of applications implemented in other countries, using the concept of Declarative Digital TV. *Adobe Flash* was used as a tool, with Object-Oriented modeling.

According to Guimaraes et. all.(2007), *Adobe Flash* has a familiar environment, providing support from the point of view of abstraction and a sense of how objects will be presented on the screen, compared to other tools that are also used in the market for developing applications for digital TV, However, one of the shortcomings of using this technology to develop TV applications is the lack of support for live editing, an important requirement to be met in Interactive Digital TV systems, since animations and interactions work using scripts (*ActionScript*).

The structure of the project can be seen in the table below:

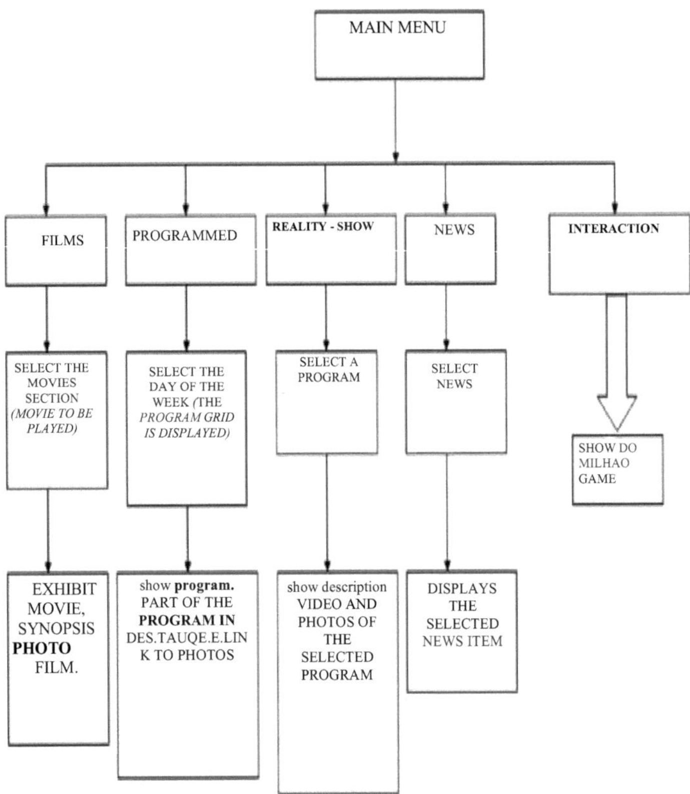

Figure 11. Structure of the system developed

To illustrate the structure of the system developed, below are some layouts of the screens that were created, simulating the Brazilian Digital TV system that is still being implemented in our country. The project documentation can be found in the Appendix.

The Home designed for the SBT television channel allows the user to view the desired programming, as well as details of favorite programs, a photo album and interactivity.

Figure 12. System Home (Channel 4 - SBT)

The Segao de Filmes Home allows the SBT television channel to make the desired movie available to the user.

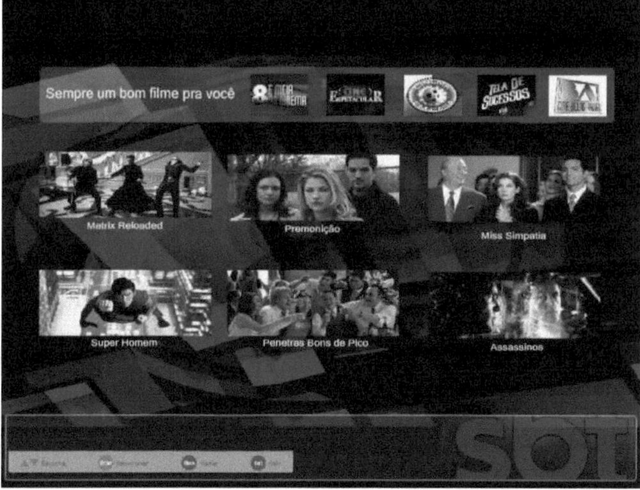

Figure 13. Home of the Movies Section (Channel 4 - SBT)

The Internal Movie Track allows the user to view details about the selected movie, as well as the synopsis, actors, director and their particularities, as well as an excerpt from the movie that will be shown later.

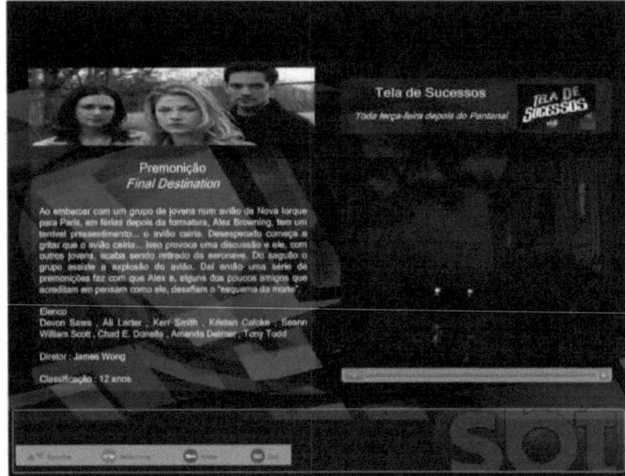

Figure 14. Internal Movie Section (Channel 4 - SBT)

The Programming Section allows the SBT television channel to display the week's program.

Figure 15. Programming section (Channel 4 - SBT)

The Internal Programming Section allows the user to view the programming for the selected day, as well as see excerpts from the program and some photos of the program.

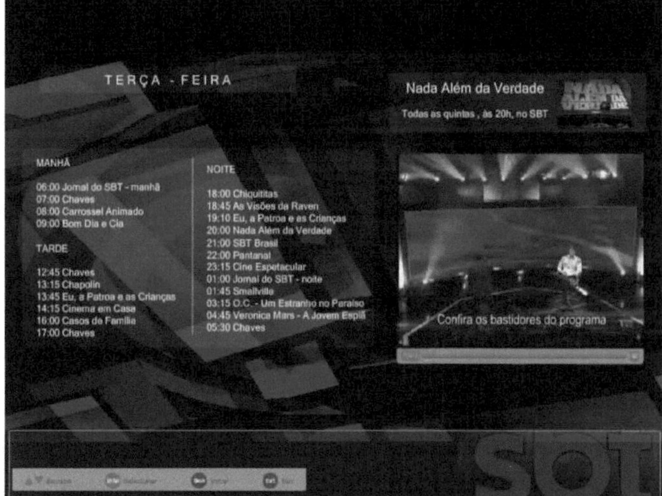

Figure 16 Internal Programming Section (Channel 4 - SBT)

Segao de Noticias provides users with the latest news, updated in real time.

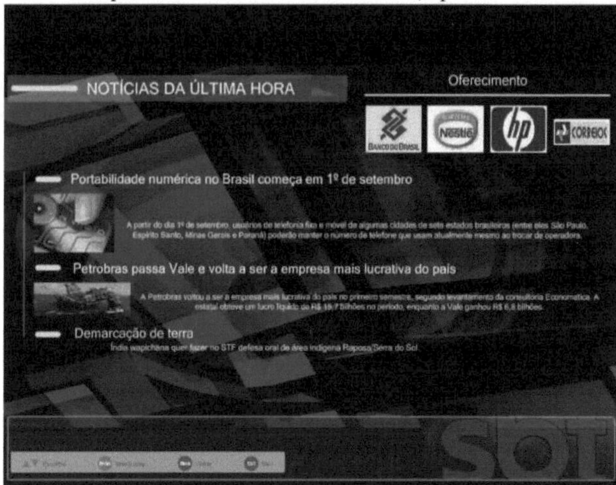

Figure 17 News Section (Channel 4 - SBT)

The Programming Section allows the GLOBO television channel to display the week's programming.

Figure 18 Programming (Channel kf5 - GLOBO).

The program section provides the user with the days of the week to choose a program.

Figure 19. Programming section (Channel 5 - GLOBO).

The Internal Programming Section allows the user to view the programming for the selected day, as well as see excerpts from the program and some photos of the program.

Figure 20 Programming section (Channel 5 - GLOBO).

The Most Viewed Videos section allows users to view the videos most viewed by viewers.

Figure 21 Most accessed videos section (Channel 5 - GLOBO).

CONCLUSION

Digital TV is being rolled out around the world, and here in Brazil it's no different. As this technology is so innovative, the social impacts will be huge, facilitating interactivity between the population and their television, such as shopping channels, participating in programs that use the population in their quizzes and reality shows. But this generates costs and is not accessible to everyone, which is why the demand for digital TV in Brazil is not being taken advantage of.

This technology will bring new changes to the job market and new IT jobs will be created. As a result, all TV stations and the job market will have to invest in their advertising and marketing departments to adapt to this new world and reach this innovative audience.

But everything that Digital TV brings you generates costs, but not just for broadcasters and the labor market, but also for your target audience, because they have to buy the devices that will transmit the signals used by Digital TV.

In this way, digital TV can only improve our daily lives, making life easier for people who are working all the time, thus promoting globalization and digital inclusion in their homes.

BIBLIOGRAPHICAL REFERENCES

ADNEWS (2008), **ATSC makes its proposal to the Brazilian government**, available at http://adnewstv.com.br/tecnologia.php?id=25532, visited in October 2008.

ANATEL, National Telecommunications Agency. Images available at: http://www.anatel.gov.br/Portal/exibirPortalInternet.do, accessed on April 28, 2008.

ARTUZI, Wison (2001), **Modulation Techniques**, available on the site HTTP://www.eletrica.ufpr.br/artuzi/cap2/pg01.html, accessed in November 2008.

BATISTA, J.C. **Economic, Technological and Social Effects of Digital TV in Brazil: alternatives for terrestrial transmission.** Texts for Discussion Series 006/2005, Rio de Janeiro, Federal University of Rio de Janeiro, 2005.

BERNAL, Volnys (2005), **Transmission of information: Multiplexing**, available at *http://www.lsi.usp.br/~volnys/courses/redes/pdf/04MULT-col.pdf,* accessed in September 2008.

BORLAND (written by the site's editor), **Maximize the Business Value of Your Software's Infrastructure and Service-Oriented Architecture (SOA).** Available on the Internet at: http://www.borland.com/br/products /middleware/index.html, accessed June 2008.

BRAIN, Marshall (2007), translated by HowStuffWorks Brasil. **How digital TV works.** Available at http://eletronicos.hsw.uol.com.br/televisao-digital.html, accessed in May 2008.

CEFET/SC (2006), **O que é TVDIGITAL**, available at, http://www.sj.cefetsc.edu.br/wiki/index.php/O_que_%C3%A9_TV_Digital%3F, visited in October 2008.

CRUZ, Renato. **Standard had better results in tests - Japanese technology was compared with American and European technology by the university Mackenzie**, January 21, 2008, article taken from the site http://txt.estado.com.br/editorias/2008/01/21/eco-1.93.4.20080121.25.1.xml.

CRUZ, Renato (2007), **Company that produces the software has already signed an agreement with manufacturers.** Report in the newspaper O Estado de Sao Paulo, September 2, 2007.

CRUZ, Renato (2008), **Japanese technology compared to American and European technology at Mackenzie University.** Report available on the internet at http://txt.estado.com.br/editorias/2008/01/21/eco-1.93.4.20080121.25.1.xml, accessed in March 2008.

DTV (2008), **Entenda a TV Digital - Official website of SBTVD - Brazilian Digital TV System**, available at http://www.dtv.org.br/index.php . Accessed in May 2008.

DBV-PROJECT, **What is the DVB Project?**,Available on the internet at : www.dvb.org, accessed in September 2008.

FUNDAQÁO CPQD, **TV por assinatura**, available at http://www.anatel.gov.br, accessed in May 2008.

FUOCO, Taís (2007). **Ministro critica a falta de celulares com transmissão de TV no País**, article in Computerworld magazine, published on November 28, 2007.

GANDELMAN, Dan Abensur. **DIGITAL TV**, published in 2/2004. Visited in September 2008. Monograph from the Federal University of Rio de Janeiro.

GUIMARÁE, Costa, SOARES. **Composer: Declarative Application Authoring Environment for Interactive Digital TV**. Monograph from the Department of Informatics - PUC-Rio, June 2007.

IDG NOW (2007), **Government studies exemption and financing for digital TV converter.** Article taken from the site at http://idgnow.uol.com.br, accessed in May 2008.

OLIVEIRA Cícero Carlos, CARVALHO Lincoln, RIBEIRO Rufino (2006). **Digital TV: international panorama and prospects for Brazil**. Monograph presented at the University of Brasília, taken from the website http://www.cic.unb.br/~bordim/TD/Arquivos/G01_Monografia.pdf, accessed on September 21, 2008.

MANHÄES M. & SHIEH P. (2005), **Canal de Interatividade: Conceitos, Potencialidades e Compromissos.** Available on the Internet at: http://www.wireles sbrasil.org/wirelessbr/colaboradores/manhaes_e_shieh/canal_de_interatividade.html.

MCLUHAN, Herbet Marshall. **The Media as Extensions of Man.** Sao Paulo. Cultrix, 1964.

MEDIALESS, **Why you shouldn't buy a digital TV abroad to work with the Brazilian standard**, from from http://www.medialess.com.br/?p=211, visited in September 2008.

MONTEZ, Carlos. **Interactive Digital TV**. 2ª . Edition. UFSC Publishing House. Florianópolis, 2005.

MOREIRA, Daniela. **USA, Europe and Japan: get to know the three digital TV standards**, February 13, 2006, article taken from the site http://idgnow.uol.com.br/telecom/2006/02/13/idgnoticia.2006-02-13.4003735509

MOREIRA, Daniela (2006), **USA, Europe and Japan: get to know the three digital TV standards.** Available at http://idgnow.uol.com.br/telecom/2006/02/13/idgnoticia.2 006-02-13.4003735509, accessed on April 28, 2008.

MOREIRA, Daniela (2007), **Digital TV: advertising will go far beyond the 30-second commercial**, available at http://idgnow.uol.com.br/telecom/ 2007/11/24/idgnoticia.2007-114.5876066434/paginador/pagina_2, accessed June 2008.

MOTA, R. & TAKASHI, Tome, **"Uma Nova Onda no Ar"**, In: Barbosa Filho, A.; Castro, C. e Tome, T (orgs.), Mídias Digitais: Convergência Tecnológica e Inclusao Social Sao Paulo, Paulinas, 2005.

MULTIMEDIA HOME PLATFORM, MHP 1.0 and GEM 1.0, available at www.mhp.org, accessed in September 2008.

Japanese broadcaster NHK , **ISDB-T.**available link http://www.nhk.or.jp/strl/publica/bt/en/pa0006 .html, visited on July 02, 2008 .

PAES Alexsandro, ANTONIAZZ Renato (2005-A), **Middleware Standards for Digital TV**. Monograph from the Master's Degree in Telecommunications at the Fluminense Federal University, taken from the website www.midiacom.uff.br/itvsoft/pdf/paes_2005a.pdf, accessed on August 15, 2008.

PAES Alexsandro, ANTONIAZZI Renato, SAADE Débora (2005-B), **MIDDLEWARE STANDARDS FOR DIGITAL TV**. Paper given at the IV Fluminense Engineering Seminar at the Fluminense Federal University, taken from the website http://www.midiacom.uff.br/itvsoft/pdf/paes_2005.pdf, accessed on August 15, 2008;

PALACIOS, Marcos Silva. **What is (really) new in Online Journalism?** Lecture given at the Faculty of Communication of UFBA, Salvador, April 2000. Available at http://www.fca.pucminas.br/jornalismocultural/m_palacios.doc, accessed on April 20, 2008.

PINHEIRO (2006), **ATSC the best standard for DIGITAL TV in Brazil**, available on the site http://www2.camara.gov.br/conheca/altosestudos/temas.html/tvdigital/TV9.pdf, visited in October 2008.

Sarmento and Reis (2007), **DIGITAL TV IN BRAZIL - A POSSIBILITY FOR SOCIAL INCLUSION IN THE AMAZON**, available at http://www.cci.unama.br/margalho/portaltcc/tcc2007/pdf/tfg022007.pdf, visited in November 2008.

Seminar of the Electronic Frontier Fundaction under the Creative Commons license BY- AS 2.5 BRASIL (2006), available on the site http://www.scribd.com/doc/275162/00559palestra-tv-digital visited in November 2008.

ROSA NETO, Antonio. **Global Attraction: the convergence of media and technology**. Sao Paulo: Makron Books, 2002.

SOUZA, Rodolfo Saboia Lima. Material Extracted from Lecture / Class (2007): **"ISDB- Brasil - Brazilian Digital TV Standard".** Monograph from the Pontifical Catholic University of Rio de Janeiro.

SOUZA, Cidcley T. de, OLIVEIRA, Carina T. de. (2005) **Specification of return channel in applications for interactive digital TV.** Available at http://www.gta.ufrj.br/~carina/artigos/iTVX.pdf, accessed in Apr. 2008.

TAGUSPARK, Instituto Superior Técnico. **DVB - Digital Video Broadcasting.** Available on the website: http://www.img.lx.it.pt/~fp/cav/ano2006_2007 /MERC/Trab_8/html%20DVBvfinal/index.html, accessed in June 2008.

TOME (2005**), COFDM is *Coded Orthogonal Frequency Division Multiple.*** *Available at* http://www.comunicacao.pro.br/setepontos/22/takashi_isdb.htm, accessed in June 2008.

UOL (2007), **From black and white to high definition TV. Check out the evolution of TV.**

Material available on the website: http://wnews.uol.com.br/site/noticias/materia_especial.php?id_secao=17&id_conteudo=342&id_coluna=7, accessed in May 2008.

W NEWS (written by the site's editorial staff), **From black and white to high-definition TV.**

Available on the Internet at : http://wnews.uol.com.br/site/noticias/materia _especial.php?id_secao=17id_conteudo=342&id_coluna=7, accessed March 2008

Printed by Books on Demand GmbH, Norderstedt / Germany